CW01266124

Series 117

This is a Ladybird Expert book, one of a series of titles for an adult readership. Written by some of the leading lights and outstanding communicators in their fields and published by one of the most trusted and well-loved names in books, the Ladybird Expert series provides clear, accessible and authoritative introductions, informed by expert opinion, to key subjects drawn from science, history and culture.

The Publisher would like to thank the following for the illustrative references for this book:
Page 43 © Kip Thorne: Keenan Pepper via Wikipedia; © Rainer Weiss: Ismael Olea via Wikipedia

Every effort has been made to ensure images are correctly attributed, however if any omission or error has been made please notify the Publisher for correction in future editions.

MICHAEL JOSEPH

UK | USA | Canada | Ireland | Australia
India | New Zealand | South Africa

Michael Joseph is part of the Penguin Random House group of companies whose addresses can be found at global.penguinrandomhouse.com

First published 2019

001

Printed in Italy by L.E.G.O. S.p.A.

A CIP catalogue record for this book is available from the British Library

ISBN: 978–0–718–18903–7

www.greenpenguin.co.uk

Gravity

Jim Al-Khalili

with illustrations by
Jeff Cummins

Ladybird Books Ltd, London

What goes up must come down

Our experience of gravity – or at least the Earth's gravitational pull – is as familiar to us as breathing. After all, we learn to cope with it before we can even talk. Every young child develops the skill and muscular strength to counter the pull of the Earth on his or her body, by first learning to sit up, then crawl, then stand up and walk. When you think about it, even the fundamental notions of 'up' and 'down' make sense only because of the direction of pull of Earth's gravity.

It is therefore not surprising that this was the first of the fundamental forces of nature that scholars attempted to explain, and we have come a very long way over the past few millennia. This book tells the story of the progress we have made in understanding what gravity is and in explaining how it fits into the laws of the Universe.

So, the first lesson is this: gravity is so much more than the simple adage 'what goes up must come down'. Stuck to the surface of our planet, we experience it merely as a force pulling us down to the ground, but we will discover that it is far richer than that – not really a 'force' at all but something altogether more profound. In fact, gravity controls the very shape of space and the passage of time and, as such, the history and destiny of the entire Universe.

Aristotelian physics

Aristotle was without doubt the greatest scientific thinker of antiquity – a man who had something to say about every aspect of the natural world. When it came to explaining why an object falls, he argued that it is because everything 'has a tendency' to move towards its 'natural' place. Of the four basic elements of which everything in the Universe was believed to be composed, the natural place for *earth* and *water* was down towards the centre of the world. Whereas the natural place for *air*, which is lighter, was thought to be above these two heavier elements, and *fire*, the fourth and lightest element, was above the air. By modern standards, of course, these ideas are not very scientific at all.

Medieval scholars in the Islamic world would improve on the Aristotelian picture. In ninth-century Baghdad, Muhammad ibn Musa was the first to suggest that celestial bodies such as the Moon and the planets were subject to the same laws of physics as on Earth – which marked a clear break from the received opinion of his day. His book *Astral Motion and the Force of Attraction* shows clear signs that he had a crude qualitative notion not too far from Newton's law of gravitation that would be laid down over eight centuries later.

But such was the Greek's influence that it wasn't until the scientific revolution in Europe in the sixteenth century, brought about by the likes of Copernicus and Galileo, that Aristotle's idea was finally rejected.

Galileo – gravity as acceleration

It has been known since the ancient Greeks that objects pick up speed as they fall. Galileo (1564–1642) discovered a mathematical formula for this. From his famous experiments rolling balls down inclined planes and carefully recording the time they took, he concluded that the distance an object drops is proportional to the square of the time it has taken. If a ball rolling down a slope covers one metre in the first second, it will have covered four metres after two seconds, nine metres after three, and so on. He also discovered that this increase in speed (the acceleration) is the same whatever the composition of the object, therefore all bodies fall at the same rate.

Galileo proposed a clever argument to support his conclusion. Think of it this way: what if a heavy body (say, a lump of iron) and a light body (say, a piece of plastic) are stuck together? If the heavy body falls faster, then it should pull the lighter one down with it, so the piece of plastic would fall faster than if it were by itself. But the combined system of both objects is even heavier than the lump of iron alone, so wouldn't it fall even *faster*? This doesn't make sense: how can adding something that falls more slowly than the iron make it fall faster? The only logical conclusion is that all bodies fall at the same rate.

Of course, Galileo is only correct if we ignore friction and air resistance. A feather will fall more slowly than a stone. The Apollo astronaut David Scott famously demonstrated Galileo's principle on the Moon in 1971 when he simultaneously dropped a hammer and a feather from each hand held at the same height. Because the Moon has no atmosphere, and hence no air resistance, the two hit the ground together.

Gravity as a universal force

By Newton's own account, it was while contemplating a falling apple on his mother's farm that he came up with his famous universal law of gravitation. In doing so, he made the connection between objects falling to the ground and the motion of the Moon around the Earth. Attributing both to one and the same force was a stroke of genius. Until then it had generally been thought that entirely different laws of nature governed the behaviour of earthly objects like falling apples and heavenly bodies like the Moon.

Newton's law states that any two objects will be attracted together by an invisible force. Thus, the Earth and the apple are both being pulled towards each other, but the distance that the Earth moves towards the apple is so extremely tiny that it could never be detected.

You may wonder how it is that the same gravitational attraction that causes the apple to fall to the ground does not pull the Moon down to Earth too? In fact, the Moon *is* constantly falling towards the Earth but in a curve that forms a near-circular orbit around it so that it never manages to get any closer. The Moon is said to be in constant 'freefall' around the Earth.

Note that Newton's law of gravity is known as a 'law' because scientists were confident that it was the last word on the subject and so they elevated its status above that of a mere scientific 'theory'. We now know that they were wrong.

The faster the cannonball is fired, the further it will travel before it hits the ground. If it were fired at an (improbable) speed of 8 km/sec horizontally from a cannon 150 km above sea level, then, were it not for air resistance slowing it down, it would remain in orbit. More correctly, it will be constantly falling towards the Earth, but the Earth curves down by the same amount as its trajectory.

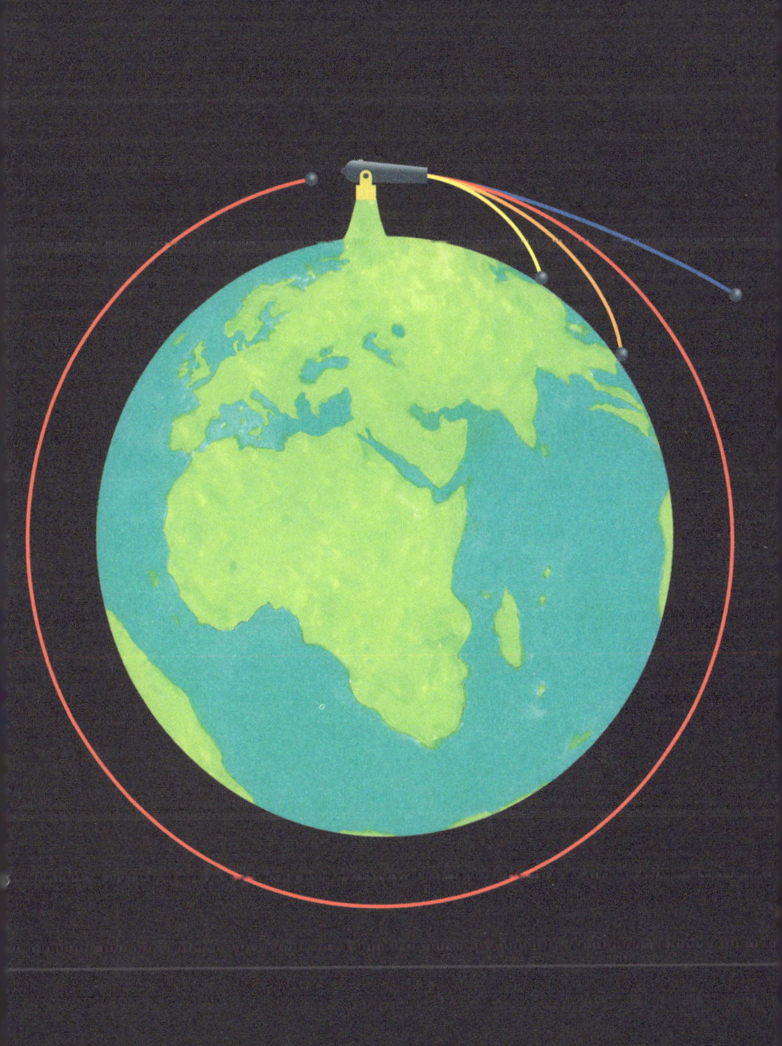

The inverse square law

Some scientific formulae are so important that they deserve an entire page to themselves in a book such as this. Newton's law of gravity is one such equation. It is written out on the page opposite and doesn't take too long to dissect and explain.

On the left-hand side of the equation, F is the force of gravity. This is the amount of 'pull' between any two objects. On the right-hand side is a group of letters denoting different quantities that together determine the strength of this gravitational force.

The two attracting bodies have masses denoted by m_1 and m_2, which are multiplied together. Thus the more massive they are, the larger the value of F (or the greater their mutual attraction).

The quantity r^2 denotes the square of the distance between the two bodies. The further apart they are, the larger this value is and hence the weaker the force of attraction. Because r^2 sits 'downstairs', this formula is known as an inverse square law.

Finally, the letter G is called the 'gravitational constant'. Newton knew it had to be there but didn't know what its value was. In fact, it was not measured until the late eighteenth century by Henry Cavendish in a remarkable experiment involving heavy lead balls hanging from a torsion wire. Today it is a number that every physics student needs to look up and plug into his or her calculations. Without it, much of modern technology would be impossible, from your bathroom scales to the Apollo Moon missions.

$F = \frac{G \times m_1 \times m_2}{r^2}$

Einstein – a new beginning

To appreciate why Einstein's concept of gravity replaced that of Newton, we must understand his new picture of space and time. In 1905, he published his special theory of relativity, in which he stated two simple principles:

1. There are no experiments we could perform that would tell us whether we were stationary or moving at constant speed. All motion is relative; and
2. Light travels at the same speed for all observers, no matter how fast they themselves are moving relative to each other.

Imagine sending a series of light pulses out into space. They will move away from you at 300,000 km/sec (the speed of light). Now get a friend to head off in a rocket alongside them at 99 per cent of the speed of light, as measured by you. You would see the light pulses only just overtaking the slightly slower rocket. But your friend still sees them overtake him at the same speed: 300,000 km/sec. This should sound astonishing. It turns out that the only way it makes sense is if your friend's time is running at a slower rate than you.

Special relativity teaches us that fast-moving observers cannot treat space and time as separate concepts but must unify them. They must view the world within 4-D spacetime, in which both spatial and temporal distances become just a matter of perspective. Thus, two people moving relative to each other will not agree on time intervals or distances between any two events, but if we combine these into 'spacetime intervals', they will always agree.

0
$\frac{1}{4}$C
$\frac{1}{2}$C
$\frac{3}{4}$C
C

The principle of equivalence

Despite its success, Einstein's special theory of relativity dealt only with objects moving at constant speeds (not accelerating). But a few years later, he had what he regarded as the happiest thought of his life: a simple yet profound connection between gravity and acceleration that went beyond what Galileo and Newton had considered. He called it his 'principle of equivalence'. It would be the breakthrough he needed for his greatest theory.

Einstein realized that the force you feel when accelerating and the force of gravity are equivalent to each other – indeed, we talk about acceleration as a G-force, where the 'G' stands for gravity. One G is the acceleration a body undergoes when falling in Earth's gravity (speeding up by 9.8 metres per second every second). Equivalently, if you are strapped in your seat inside a rocket out in empty space that is accelerating at one G, this would feel no different to you than if the rocket were still sitting on the launch pad on Earth and your seat was facing straight up with gravity pulling you down.

When a body is in freefall in Earth's gravity, it feels no forces acting on it. The aptly named 'vomit comet' (or, more correctly, the 'reduced gravity aircraft') shows this equivalence dramatically. Used to give astronauts the sensation of weightlessness they will feel when in deep space, it follows a parabolic flight path during which both the aircraft and its occupants are in freefall for a short while.

Stephen Hawking taking a trip in the 'Vomit Comet' in April, 2007.

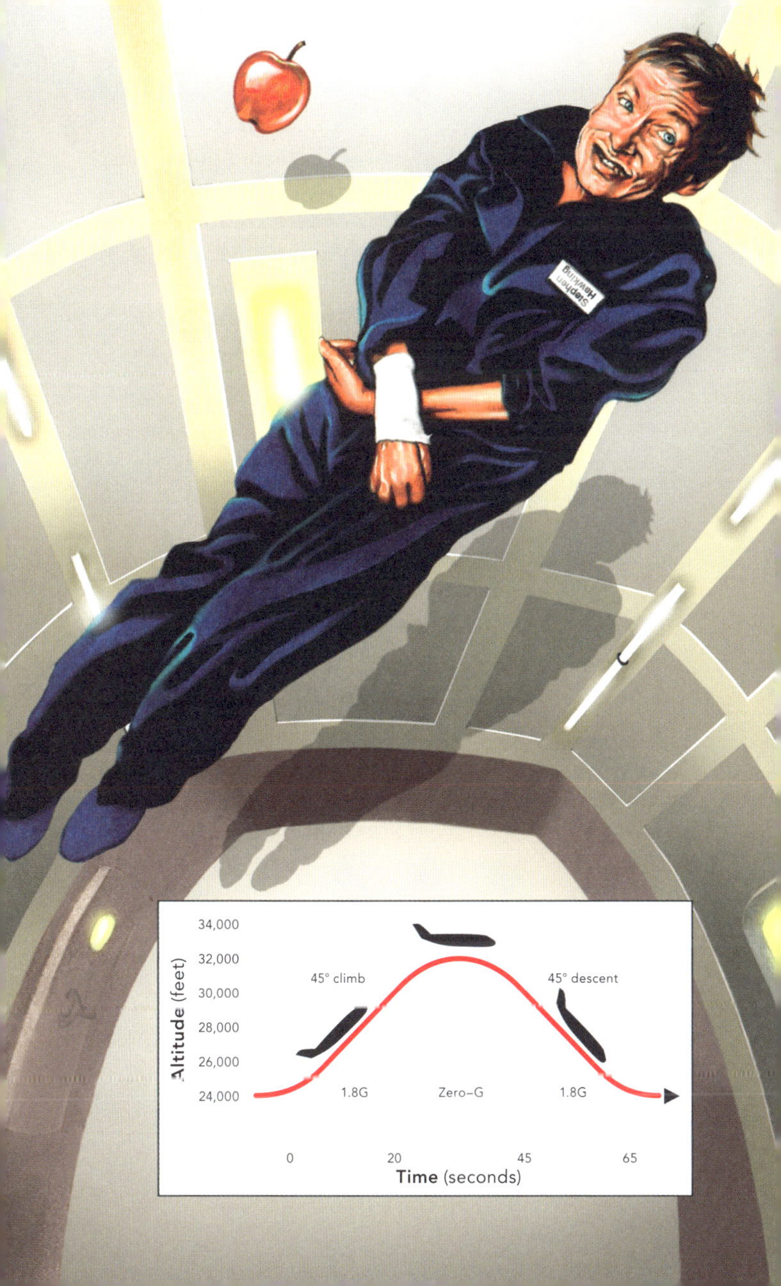

Stephen Hawking
Altitude (feet)
34,000
32,000
30,000
28,000
26,000
24,000
45° climb
45° descent
1.8G
Zero–G
1.8G
0
20
45
65
Time (seconds)

Gravity as the shape of spacetime

Newton defined gravity as a force that every object exerts on its surroundings – it's the invisible 'rubber band' that pulls masses together. Einstein gave a deeper and more profound explanation than this. He said that gravity is the shape of space itself. Any mass bends, or warps, space around it, which in turn causes objects in that space to move differently from the way they would move if the mass were not there and the space were flat.

More correctly, since the lesson of Einstein's special theory of relativity (page 14) is that space and time must be unified into 4-D spacetime, masses in fact cause *spacetime* around them to curve. This is very difficult to visualize so it's simpler to think of four-dimensional spacetime squashed down to a two-dimensional sheet so that we can imagine it stretching and bending.

Imagine rolling a small ball across an empty trampoline. It will move in a straight line. But if someone is standing in the middle of the trampoline and the ball is rolled again, you will see that its path now deviates from a straight line, bending around the dip. From a bird's-eye view above the trampoline it would appear as though a mysterious attractive force were being exerted on the ball by the person standing in the middle, causing it to be attracted towards them.

Newtonian gravity is the bird's-eye view, whereas Einsteinian gravity explains the motion in terms of the shape of the trampoline (spacetime) itself.

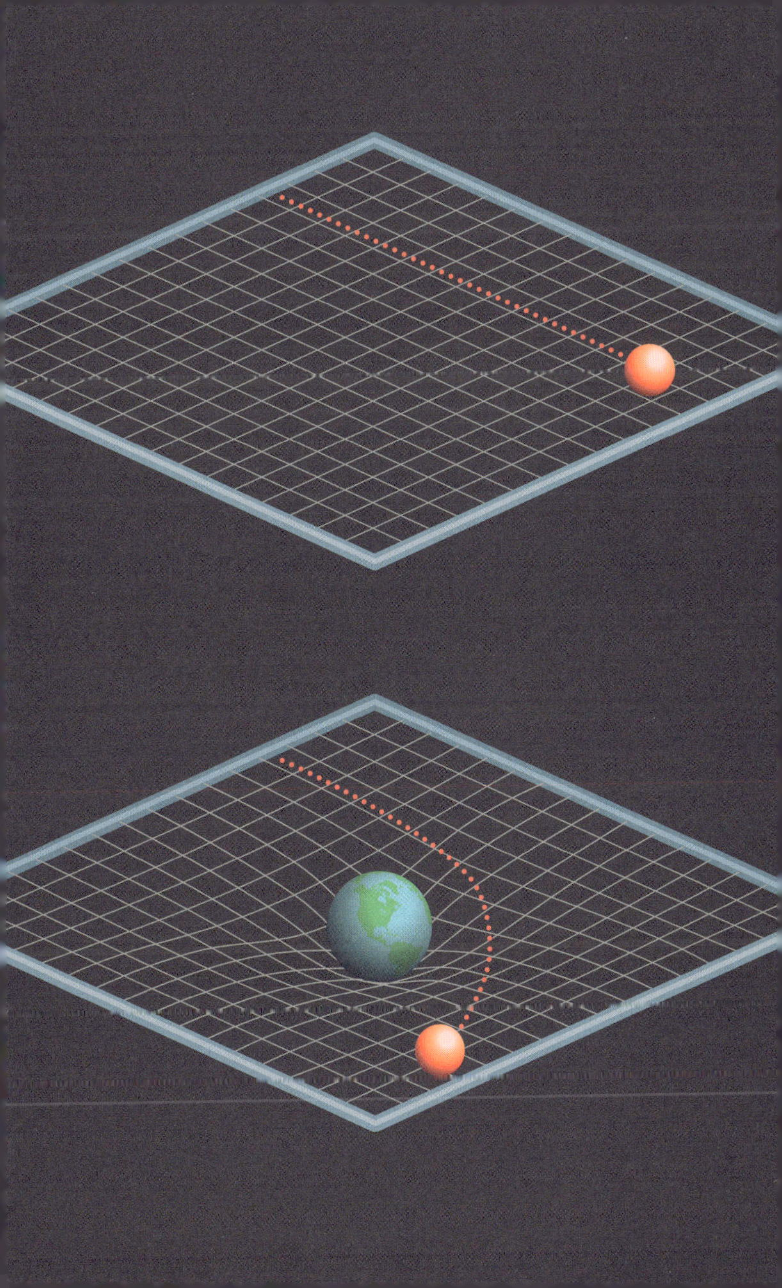

Cosmology

Einstein's new theory of gravity was completed in 1915 and is today known as the general theory of relativity (or simply general relativity) – to distinguish it from his special theory of relativity published ten years earlier, which described the 'special case' of how objects move in the *absence* of gravity or acceleration. General relativity is encapsulated in Einstein's mathematically beautiful field equations, which provide a deep connection between energy, matter, space and time, and are far more complicated to dissect than the equation Newton wrote down 230 years earlier.

Soon after Einstein published his theory, he and others began applying it not only to individual masses like the Sun and the planets but to the entire Universe, thus extending Newton's grand project, and so the modern field of cosmology was born: the study of the size, shape, birth, evolution and fate of the Universe.

A major prediction of general relativity is that the Universe is expanding. This was confirmed in 1929 by the astronomer Edwin Hubble, who observed that at the vastest cosmic scales, galaxies are flying apart from each other, not because they are moving through space but because the space in between them is stretching.

Most of the ideas, discoveries and advances discussed in the rest of this book arise out of general relativity – Einstein's theory of gravity, space and time. But despite the many successes it has had over the past century, a number of mysteries remain, as we shall also see.

$$R_{\mu\nu} - \frac{1}{2} R g_{\mu\nu} = \frac{8\pi G}{c^4} T_{\mu\nu}$$

Expanding into what?

Before moving on, it is worth dwelling for a moment on a few of the trickier concepts in cosmology. For example, if the Universe is expanding, what is it expanding into? This gets even more confusing when we consider that the Universe might be infinite in extent. How can something that already goes on forever get bigger? Surely there is no more room for it. The answer, whether it helps you or not (!), is that the Universe contains the entirety of space and time and so it doesn't require any further space 'outside' it to expand into.

Looking at a galaxy ten billion light years away we are seeing it as it was ten billion years ago, because the light from there has taken that long to reach us. Beyond the edge of the visible universe is a region whose light has not reached us yet and never will, because it is expanding away from us so fast. But this visible limit is not the entirety of the Universe, which does not have an edge, just as the surface of the Earth has no edge.

Another confusion is how all distant galaxies are receding away from us equally in every direction. Surely this means we must occupy a privileged position at the centre of the Universe. This is not correct. Think of us as a point somewhere on the surface of a balloon. As you blow the balloon up, the rubber stretches and all points around us move away equally. We would see the same thing happening wherever we were on the balloon.

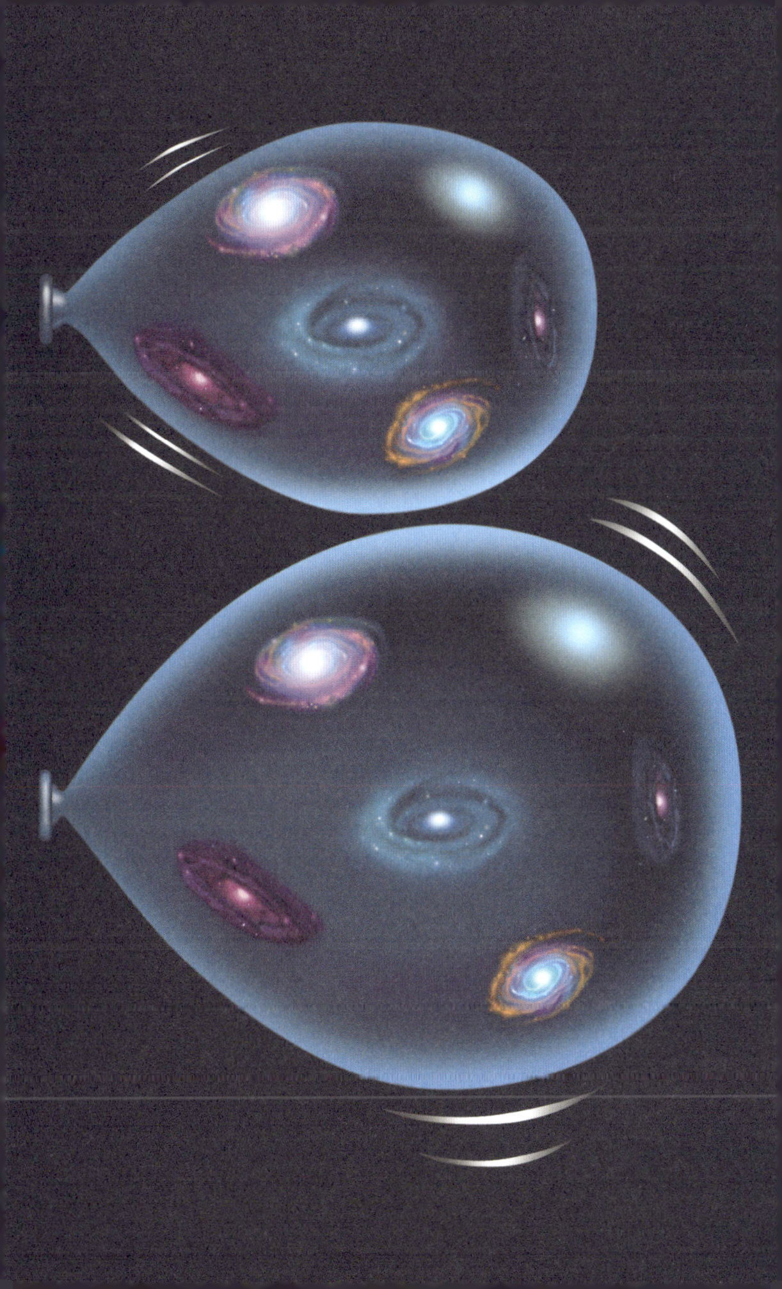

The Big Bang

The cumulative gravitational pull (to use the old Newtonian notion for simplicity) of all the matter in the Universe should be slowing the expansion down, so what started it in the first place? This initial 'kick' that got the Universe going is what we refer to as the Big Bang. It is the event that marked the birth of space and time and everything within it.

The expansion of the Universe is the most persuasive evidence for the Big Bang. If we could run time backwards, we would see the Universe shrinking. And if we could go back far enough into the past, we would see all matter coming together, squeezing into a smaller and smaller volume, until we reach the moment of the Universe's birth 13.8 billion years ago.

The Big Bang model is also supported by two other crucial observations. The first is that it accurately predicts how particles came together to make atoms and how the lightest chemical elements were formed. And the second is that the temperature of deep space today is exactly what the theory predicts if the Universe has been cooling down for the past 13.8 billion years. This afterglow of the Big Bang is called the cosmic microwave background.

However, there is no point in the Universe where we could say 'the Big Bang happened here'. In a sense the Big Bang happened everywhere, since everywhere was in the same place then.

What 'caused' the Big Bang to happen is another issue entirely, and is a question cosmologists are still working on.

BIG
BANG!
ABSOLUTE TOSH
Fred Hoyle

Black holes

The first scientist to mathematically study the properties of a gravitational field around a large concentration of mass was Karl Schwarzschild in 1917. But it wasn't until the mid 1960s that the phrase 'black hole' was coined to describe such an object.

Schwarzschild said that when a massive enough body is crushed by its own gravity, it will squeeze to a critical size beyond which there is nothing to stop it from collapsing further. This is defined today by what is called the 'event horizon' – the spherical boundary that marks the point of no return. Outside the horizon, gravity is strong but finite, and it is possible for objects to escape its pull. But within the horizon, nothing can escape, not even light.

When a large star explodes as a supernova, its core can collapse under its own weight to form a neutron star, an object so dense that a teaspoon of its matter has a mass of about five billion tons. And if the star's core continues to shrink through its own event horizon, it will form a black hole and its entire mass collapses to a single point called 'the singularity'.

A black hole formed by the collapse of a star ten times the mass of the Sun will have an event horizon that is thirty kilometres across (as viewed from outside). But if the Earth could be squeezed hard enough to turn into a black hole, its event horizon would be the size of a pea.

Falling through an event horizon

If you were to watch, from a safe distance, an unfortunate astronaut plummeting into a black hole, he would appear to slow down as he fell until he finally stops, frozen, just outside the event horizon. This is because time itself slows down in a gravitational field (see page 32). But his image would very quickly fade away, not because you have 'seen' him fall through the horizon but because the wavelength of the light reaching you has shifted beyond the visible.

For the astronaut, things are different. He is in freefall towards the black hole, speeding up all the time. He won't be aware of the moment he passes through the event horizon, and it doesn't suddenly get dark for him, because light can stream in from outside. He can see out, but you can't see in. However, the light he sees will be focused into a shrinking bright patch, like the view of the receding entrance from inside a dark tunnel.

Inside the black hole, space and time are so warped that the distance from the event horizon to the singularity is now a direction in time. This interchange of space and time explains why any object falling into a black hole has no choice but to move towards the singularity – no more than we can avoid moving towards the future.

But the poor astronaut won't even be able to enjoy the view because his body will be stretched like spaghetti by the gravitational field. This is known technically as 'spaghettification'.

Are black holes real?

The quick answer is yes. And there are even different types of black hole.

Stellar black holes are formed when massive stars collapse. Although they cannot be seen, they are detected indirectly from the way they influence visible matter nearby. Most stars come in pairs, or binary systems, that orbit around each other. If one of the pair collapses into a black hole, its gravitational effect will remain the same, so we see the motion of the remaining visible star as it continues to orbit its invisible partner. The black hole can also suck material from its partner and this gas will spiral in towards the hole's event horizon, heating up as it does so and forming an 'accretion disc'. Just before it falls through the horizon, the ultra-hot gas will give off powerful X-ray bursts. The first black hole ever detected is part of the binary system Cygnus X-1, which is about 6,000 light years from Earth.

We now know that supermassive black holes exist at the cores of large galaxies, including our own; called Sagittarius A*, ours has a mass about four million times that of our Sun. Scientists think that such huge black holes would have formed from the accumulation of vast amounts of compact stellar gas in the dense centres of galaxies. They can capture and swallow any nearby star that ventures too close.

It has also been suggested that mini black holes, created just after the Big Bang, may exist, smaller than an atom but with as much mass as a large mountain. If they do exist, then by now they will have completely evaporated away and should ultimately explode in a tremendous final explosion of high energy radiation, which astronomers are looking out for.

What a black hole may look like. The part of the accretion disc behind the black hole is visible above and below the black hole due to gravitational lensing (see page 38).

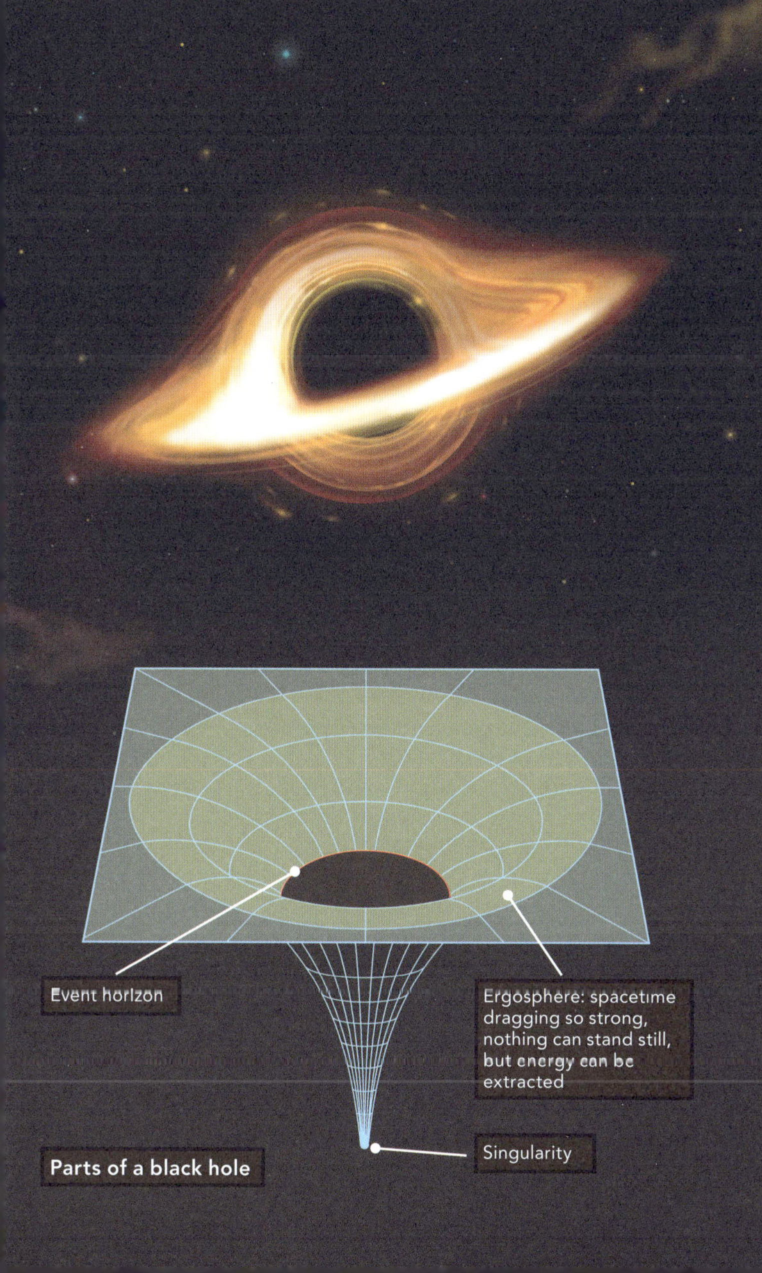

Parts of a black hole

Gravity's effect on time

General relativity states that the gravitational field is the shape of spacetime, so gravity will also have an effect on time; the greater the mass of a body, the stronger its gravitational field and the more it slows down the flow of time. The slowing of time due to the Earth's gravity was confirmed in a famous experiment carried out by Robert Pound and Glen Rebka in 1960, which involved measuring the tiny difference in frequency of light given off by radioactive atoms at the top and bottom of a tower.

The higher in altitude we go, the weaker the Earth's gravitational pull is and hence the faster time can run. In fact, if your watch is running slow, hold it above your head and it will speed up since it is experiencing weaker gravity. (The effect in this case will be very tiny indeed.)

The effect of gravity on time is routinely taken into account on board GPS satellites. Without corrections for the changes in the rate of flow of time between the satellites and the ground, we would not be able to fix our position with our smartphones and satnavs to the accuracy of a few metres, as we have become used to. Ignoring Einstein would mean that satellites would get our position wrong by over ten kilometres per day.

When gravity is very strong, such as close to a black hole, spacetime is curved dramatically and time runs much more slowly. This suggests that by orbiting a black hole you will time travel into the future, since on your return to Earth you would find everyone else has aged more than you.

07 12

Earth's gravitational field

Our planet is not a perfect sphere, nor is its mass uniformly distributed throughout its volume. This means that there are variations in the gravitational field around its surface. By measuring these gravity anomalies from space, scientists can study how mass is distributed around the planet and how it varies over time. This data is important for studying Earth's geology, oceans and climate. It can help with tracking the flow of magma inside the planet, the circulation of ocean currents, the thinning of polar ice sheets and hence understanding the causes of rising sea levels.

The Gravity Recovery and Climate Experiment (GRACE) is an Earth-monitoring satellite mission that uses a microwave ranging system to accurately measure changes in the distance between two identical spacecraft orbiting the Earth about 220 km apart. The system is sensitive enough to detect separation changes as small as a tenth of the width of a human hair. As the satellites circle the globe fifteen times a day, they sense minute variations in Earth's gravitational pull, making them, independently, speed up or slow down very slightly, causing the distance between the satellites to change. By monitoring this changing distance and combining it with GPS data, a detailed and strangely lumpy map of the Earth is produced.

But the Earth's own gravitational field is not the only thing that affects it. The combined influence of the Moon and Sun's gravity, along with the rotation of the Earth, are responsible for the tides – the periodic pulling and squeezing of the Earth's oceans.

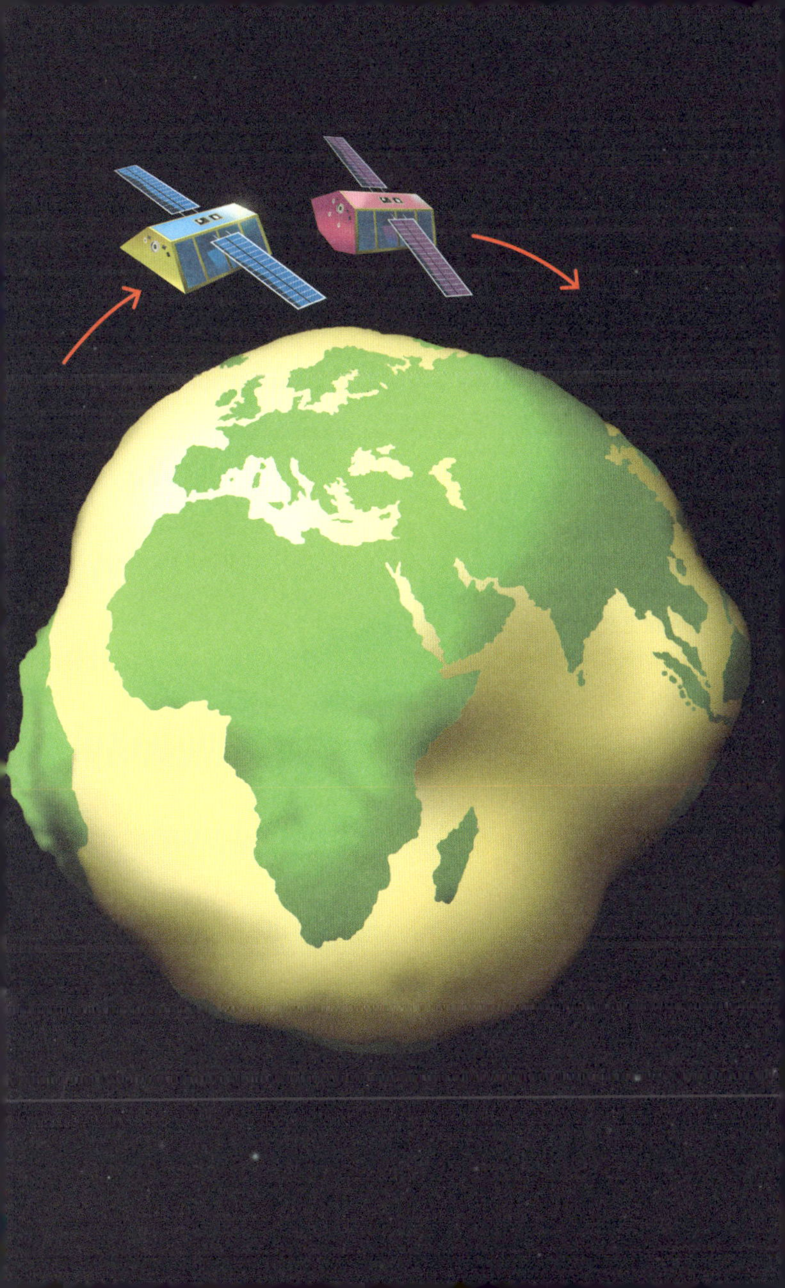

Gravity in our Solar System

Gravity has shaped everything in the Universe. Matter clumps together under its influence, forming all the galaxies and stars. Our Sun was born about 4.6 billion years ago from the collapse of a giant cloud of hydrogen and helium gas, along with heavier elements thrown out by nearby supernovae. Like all the other planets in the Solar System, the Earth was formed from the leftover material orbiting the young Sun.

The reigning hypothesis for the Moon's formation is that a Mars-sized body collided with Earth. Named 'Theia', after the mother of the Moon goddess Selene in Greek mythology, it would have caused vaporized chunks of the Earth's crust to be thrown into space, which eventually clumped together to create our Moon. A possible refinement to this hypothesis suggests that multiple objects struck Earth, and the debris from these collisions merged over time.

Five billion years from now the Sun will expand to become a red giant. Its core will collapse and heat up while its outer layers will expand outwards until it swallows Mercury. It will fill half the sky and become much hotter and brighter than it is today. After a further billion years, it will shed its outer layers to form a ring of gas called a planetary nebula, at the heart of which will sit its dying core: a white dwarf star the size of the Earth comprising mainly crystallized carbon and oxygen. Anything still living at this time will have had to find a new home in a different star system.

Gravitational lensing

General relativity predicts that the Sun's gravity will bend the path of light coming from distant stars hidden behind it. However, this can only be tested when the Moon blocks out the light from the Sun itself. During the total solar eclipse of 1919, the astrophysicist Arthur Eddington did just that. He led an expedition to the Amazonian jungle where he saw that stars in the line of sight of the Sun appeared to have shifted slightly in their position relative to where they were in the night sky without the influence of the Sun's gravity.

This effect is part of a more general phenomenon called gravitational lensing whereby light can be bent or focused in ways similar to the action of a lens in optics, only here it is the shape of spacetime itself that influences the path of the light.

Gravitational lensing came of age in 1979 when an interesting feature of a quasar nine billion light years away was observed. Quasars are very distant, very bright, active galactic cores. The warped spacetime around a nearer galaxy sitting directly in front of the quasar split the light coming from it into two paths, so that it appeared as a double image.

Gravitational lensing is today established as a very useful tool in astronomy. It has helped us understand the way matter is distributed in the Universe, as well as the structure and properties of galaxies and the gravitational nature of dark matter (see page 42). It has also thrown up a whole host of interesting astronomical phenomena, such as so-called Einstein rings, galactic microlensing and giant luminous arcs.

Gravitational waves

General relativity predicts that matter not only bends spacetime but causes it to undulate when disturbed, creating waves that travel radially outwards from the point of disturbance at the speed of light, squeezing and stretching spacetime as they pass through. It took a hundred years for this prediction to be confirmed.

In a distant galaxy over one billion years ago, two black holes swirled ever closer around each other until they finally collided and merged with incredible violence. In that final fraction of a second, the resulting disturbance of spacetime sent gravitational waves washing across the Universe, finally passing through the Earth on the morning of 14 September 2015, where they were detected by jubilant scientists.

Each of the two identical LIGO (Laser Interferometer Gravitational-Wave Observatory) facilities in the US, which picked up the gravitational waves signal, is an enormous L-shaped construction in which laser beams are sent along the two 4-kilometre arms to measure, very accurately, the temporary tiny differences in their lengths caused by the gravitational waves.

In 2017, gravitational waves from the merger of two neutron stars were detected on Earth. This powerful event, known as a kilonova, took place 130 million light years away and is already regarded as one of the most important astrophysical events ever seen.

Until the discovery of gravitational waves astronomers had only ever been able to study the Universe by capturing its light. With this new tool – gravitational wave detection – a new era in astronomy is beginning.

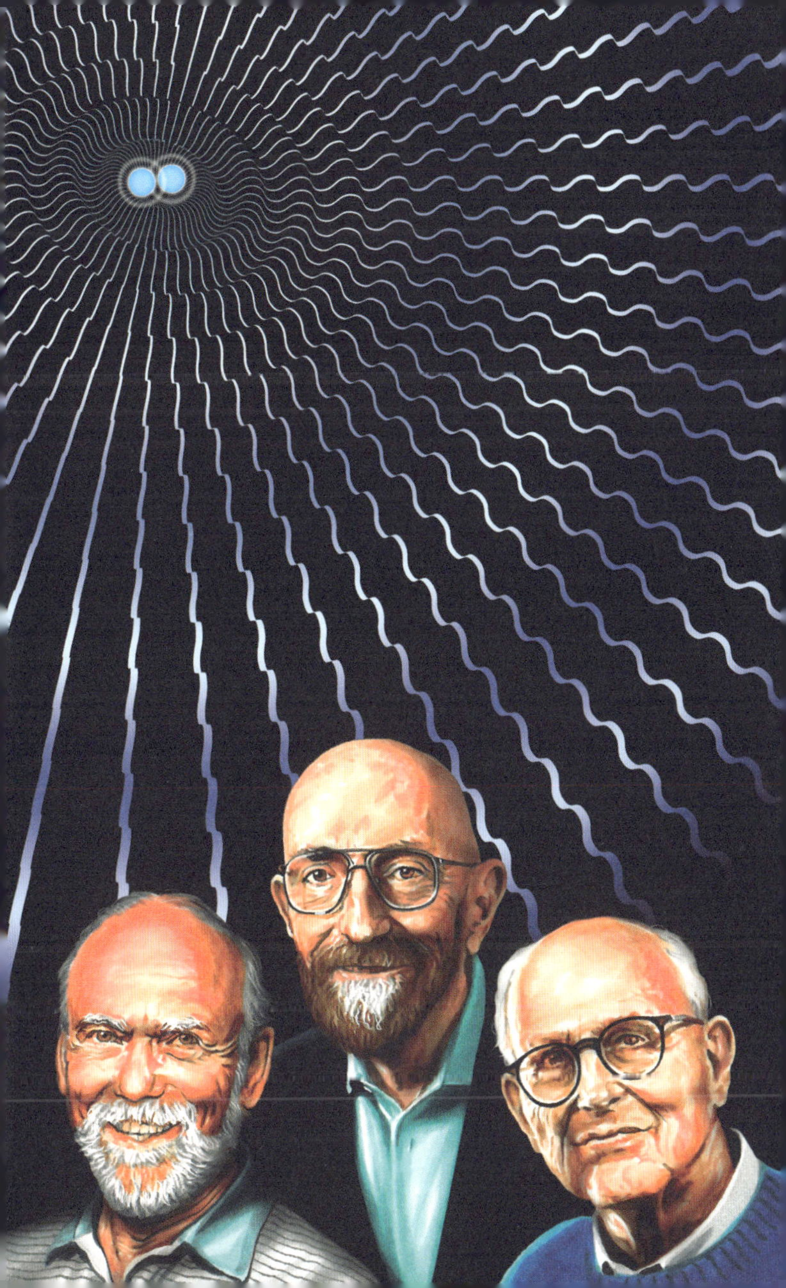

Dark matter

In 1933, the Swiss astrophysicist Fritz Zwicky was studying the motion of galaxies around the outer edge of the giant Coma galaxy cluster and, because of the additional gravity necessary to describe the way they moved, estimated that there had to be 400 times (an over-estimation) as much matter in the cluster than could be accounted for. He called this invisible stuff '*dunkle Materie*' ('dark matter').

Vera Rubin provided further evidence for dark matter in the 1970s when she observed that the spiral arms of galaxies were spinning too fast if all that was holding them in place was the gravitational attraction of the visible galactic contents (stars and interstellar dust and gas). She estimated that most galaxies must contain roughly six times as much dark matter as normal matter.

For the past four decades the hunt has been on to find out what dark matter is made of. It remains, at the time of writing, an unsolved problem in astronomy. So, what can we say about it so far? It is known that dark matter cannot be composed of atoms, or the particles that make up atoms, otherwise it would feel the electromagnetic force and hence emit and absorb light. It is therefore thought to consist of a new, as yet undiscovered, substance that interacts with normal matter only gravitationally.

Although most astronomers accept the existence of dark matter, the fact that we still do not know what it is made of is a growing source of frustration.

SPIRAL

Dark energy

In 1927, the Belgian priest and theoretical physicist Georges Lemaître predicted the expansion of the Universe from general relativity, a result subsequently confirmed by the observations of Edwin Hubble.

But the question that then occupied cosmologists was whether this expansion would continue forever or whether it would stop and possibly even go into reverse if the gravitational pull of everything in the Universe were to cause it to collapse in on itself. It all depended on how much mass the Universe contained.

In 1998, astronomers studying light coming from distant galaxies measured the speeds at which those galaxies were receding from us due to the expansion of the Universe and made a shocking discovery. They found them to be moving away more slowly than they should have been based on how far away they were. Since the light reaching us from these galaxies set off when the Universe was much younger, their slower-than-expected speed indicated a slower expansion rate in the past. This meant that the rate of expansion of the Universe today is greater than it was in the past. So, rather than gravity putting the brakes on the expansion of the Universe, something else was speeding it up.

The only way for this to be possible is if some mysterious repulsive force is counteracting gravity and stretching space ever more quickly. For want of a better name, it is called 'dark energy' – not to be confused with dark matter. According to current understanding, this dark energy will result ultimately in what is called the 'heat death' of the Universe many billions of years from now.

HEAT DEATH

A MYSTERIOUS REPULSIVE

FORCE

IS DRIVING THE EXPANSION
OF THE UNIVERSE

Wormholes

Einstein was bothered by the notion of a singularity at the centre of a black hole because general relativity predicts this to be a point of zero size. So, by using a mathematical trick, he showed, together with his collaborator Nathan Rosen, that the interior of a black hole might not contain a singularity but instead be a bridge to a parallel universe.

However, such an 'Einstein-Rosen bridge', as it came to be known, could never be used as a practical means of getting across to a neighbouring universe because it would be unstable, and in any case the black holes at either end would have event horizons blocking any escape.

In the mid 1950s, the American physicist John Wheeler showed that a tunnel in spacetime could bend round to provide a shortcut joining two different regions of our Universe together (like the handle on a coffee mug), for which he coined the term 'wormhole'. Later, hypothetical stable, 'traversable' wormholes were postulated, which did not have event horizons at their ends.

In the late 1980s, it was pointed out that if wormholes really existed then they could be used as time machines because their mouths would not only link two different regions of space but two different times. This would allow for the bizarre possibility of you travelling through a wormhole only to come out the other end before you entered.

A recent intriguing idea is that miniature wormholes on the quantum scale might account for the way two entangled particles are linked across vast distances.

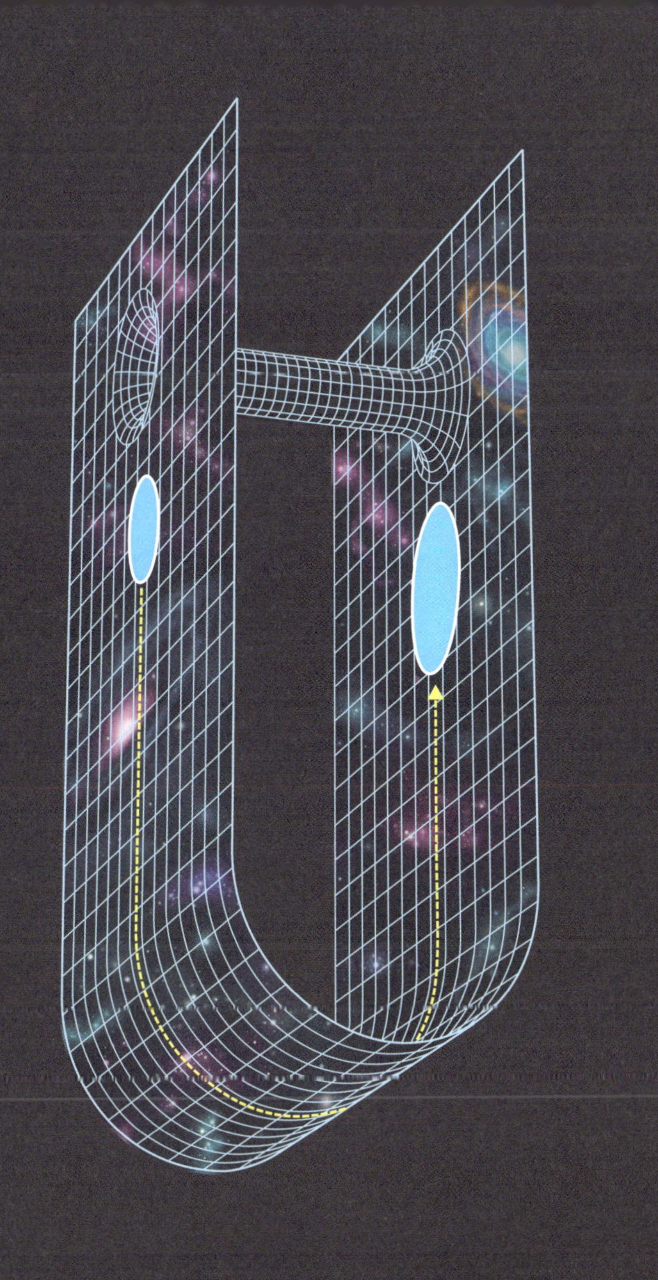

Quantum gravity

One of the biggest challenges in science today concerns reconciling the two pillars of twentieth-century physics: general relativity and quantum mechanics. While the former describes space and time as continuous and infinitely divisible, quantum mechanics is 'lumpy' and probabilistic and describes matter in terms of elementary particles and their interactions. A unification of these two successful yet incompatible descriptions of reality, an all-encompassing theory of 'quantum gravity', is the holy grail for many physicists.

There are currently two leading contenders for a theory of quantum gravity. 'String theory' begins with the quantum mechanical properties of matter within spacetime. Its basic idea is that all elementary point-like particles are in fact tiny vibrating strings and it requires there to be several higher dimensions tightly curled up and currently undetectable. However, despite much theoretical effort over almost half a century, string theory has yet to deliver.

The other approach, called 'loop quantum gravity', starts from general relativity and assumes that spacetime itself, rather than the matter it contains, is the more fundamental concept. Quantizing spacetime suggests there is a smallest length and shortest time that cannot be subdivided. The loops in the name refer to the closed geometric paths that link adjacent quanta of space. It is the nature of these loops that determines the shape of spacetime.

Time will tell whether string theory or loop quantum gravity is correct, or if indeed neither is and an entirely new approach is needed to merge the quantum world with Einstein's gravity.

21ST CENTURY PHYSICS

Why is gravity so weak?

Of the four forces of nature – gravity, electromagnetism and the strong and weak nuclear forces – gravity is by far the weakest, despite its dominant role in shaping the Universe on the cosmic scale. So, to unify all the forces under one banner – one 'theory of everything' – we will have to understand *why* gravity is so feeble.

Strange as this sounds, it may be because some of gravity's strength is leaking away into dimensions beyond the familiar three of space and one of time that we experience. Indeed, ours may well be but one of many universes – each a bubble floating in a higher-dimensional space called the 'multiverse'.

It is possible that we might one day find experimental evidence for these other dimensions. If they do exist, they might help us answer a number of deep questions, such as how our universe came into existence in the first place and why its physical properties are so perfectly tuned for stars, planets, atoms, molecules and even life itself to exist.

Some theorists suggest that a hypothetical particle called the graviton is associated with the gravitational field in the same way that the photon is the particle of the electromagnetic field. According to this picture, while the other three forces remain trapped within our four-dimensional spacetime, gravity is free to roam across the multiverse. Gravitons can bleed out of our universe, therefore weakening the strength of gravity that we experience.

As you can see, there are still plenty more problems to solve before we fully understand gravity.

CREATE YOUR OWN
MULTI-VERSE!
GRAVITON

Further reading

Jim Al-Khalili *Black Holes, Wormholes and Time Machines* (Taylor & Francis, 1999)

Jim Baggott *Mass: The Quest to Understand Matter from Greek Atoms to Quantum Fields* (Oxford University Press, 2017)

Bill Bryson *A Short History of Nearly Everything* (Transworld, 2003)

Marcus Chown *Big Bang: A Ladybird Expert Book: Discover How the Universe Began* (Michael Joseph, 2018)

Marcus Chown *The Ascent of Gravity: The Quest to Understand the Force that Explains Everything* (Weidenfeld & Nicolson, 2017)

Timothy Clifton *Gravity: A Very Short Introduction* (Oxford University Press, 2017)

Neil Degrasse Tyson *Astrophysics for People in a Hurry* (W. W. Norton & Company, 2017)

Albert Einstein *Relativity: The Special and General Theory* (Project Gutenberg)

Brian Greene *The Elegant Universe: Superstrings, Hidden Dimensions and the Quest for the Ultimate Theory* (Vintage, 2000)

Lisa Randall *Dark Matter and the Dinosaurs: The Astounding Interconnectedness of the Universe* (Ecco Press, 2015)

Govert Schilling *Ripples in Spacetime: Einstein, Gravitational Waves, and the Future of Astronomy* (Harvard University Press, 2017)

Simon Singh *Big Bang* (Harper Collins, 2004)